T0173243

2nd EDITION

Reading Anthology
Foundation

The pond

I spider, I duck, I butterfly,
How many tadpoles do you spy?

Look quickly at the groups you see.
Which have 9? Which have 3?

10 crayons on the bed, 10 teddies in the bed!
Is there the same number of books to be read?

Across the room are zooming cars.
Count up and write how many
there are!

Jack trots along behind his sister Jill,
Over the river and up the big hill.

The men on the water, inside their small boat,
Are near some green frogs on a log where they float.

What number of sheep at the farm can you spy?
Is it more than the number of birds in the sky?

Look at the fields. What's the number of trees?
Is it smaller than 9? Is it bigger than 3?

The castle is standing opposite town. The cart goes on the path between them, up and down!

High above the castle, how many
flags show?
Are there fewer than the number of
windows below?

Minibeasts

How many caterpillars can you spy?
How many worms and butterflies?

Which groups can you see have a total of 4?
Which groups have fewer?
Which groups have more?

Necklaces

See the necklaces these beads make,
And all the beads' colours, sizes
and shapes.

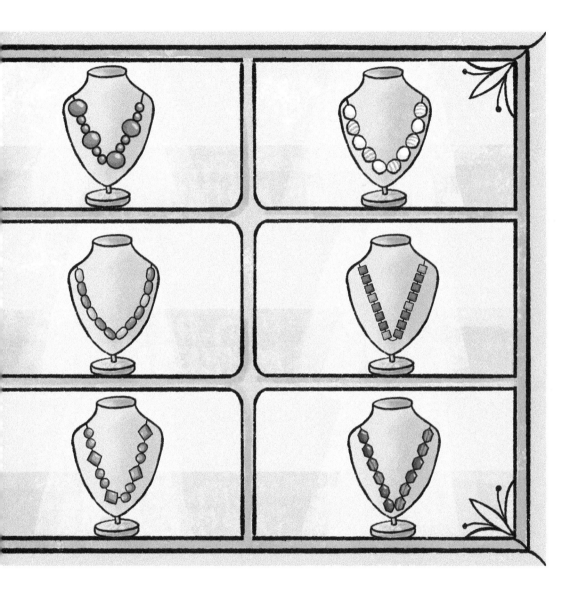

Each necklace has a pattern of beads. To make one longer, which beads would you need?

What different things can circles be?
How many rectangles can you see?

Find a triangle, find a square.
Are there any more shapes there?

3D shapes

Cylinders, cones, spheres and cubes,
pyramids and cuboids, too.

Find shapes that roll and shapes that stack. Did you spot a few?

Jenny's classroom

Ted keeps his bag behind him, on the corner of his chair.

Kim leans on the edge of the table with his hand up in the air.

The bookshelf's near the toy box.
Outside of the window's a tree.

The crayons are under my best
friend's hand — and next to her is me!

The zoo

Which is the tallest giraffe in the zoo?
Which is the heavier kangaroo?
Look for the shortest crocodile, too!

Which lion has the widest mane?
Which pathway is the narrowest lane?
Which animals are all the same?

The see-saw

Look at the animals on the see-saw!
Which weigh less, which weigh more?

How many animals came to play?
Who joined the fun? Who ran away?

These are the vases in Yasmine's store.
Some hold less and some hold more.

Which vases are empty in the shop?
Which are full, right to the top?

Sven's day at school

At 9 o' clock, where is Sven?
Can you see what he did then?

At lunch, what did the clock face say?
What happened last, to end the day?

The toy cupboard

Look at the toys in this display!
Why have groups been made
this way?

Do some toys fly? Do any float?
Find the spheres, then find a boat.

Numbers 0 to 10

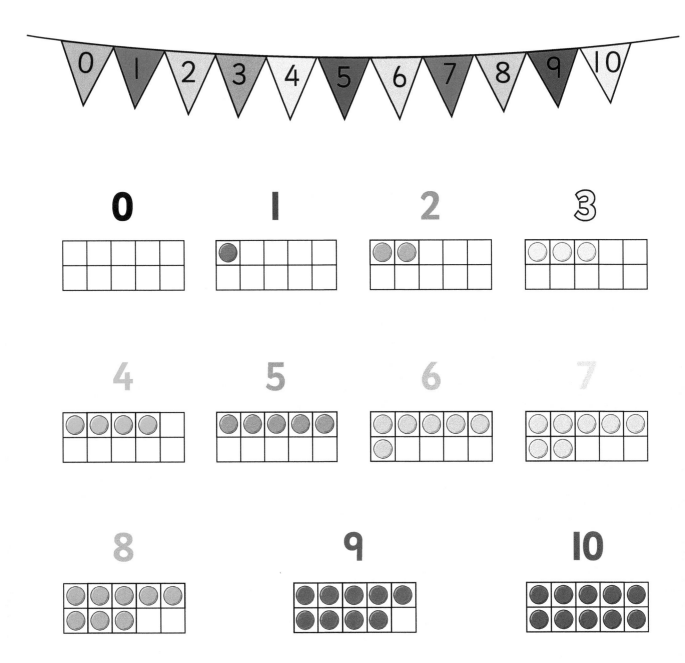

0 1 2 3 4 5 6 7 8 9 10

0   **1**   **2**   **3**

4   **5**   **6**   **7**

8   **9**   **10**